The Earth Pledge Book

*****

A Call for Commitment

*****

Compiled And Introduced By

John F. Ince

The Earth Pledge Book
❋ *2* ❋

Copyright 1994 by John F. Ince

All rights reserved. Written permission must be secured from the publisher to use or reproduce any part of this book except for brief quotations in critical reviews or articles.

Published in Sausalito, California by Timely Visions Publishing Company, P.O. Box 2542, Sausalito, CA 94966; Distributed by SCB Distributors, Gardenia, CA 90248-2129

Cover design by Madelein Fishman and Alan Guansing
Editorial and layout assistance by Shema Satya

Library of Congress Cataloging-in-Publication Data
The Earth Pledge Book: A Call To Commitment To Save Our Environment
Includes 111 Quotations by Famous Environmentalists And 111 Simple Earth Pledges
1. Environment 2. Nature 3. Education
I. Ince, John F. 1948-   Title: The Earth Pledge Book
Library of Congress Catalogue Card Number: 95-90260

ISBN 0-934239-25-8

Printed on Recycled Paper With Soy Based Ink By Patterson Printing, Benton Harbor, MI 49022

Table of Contents

❋ The Earth Pledge - Page 5
❋ The Objectives of the Earth Pledge Campaign - Page 6
❋ Board of Advisors of The Earth Pledge Campaign - Page 8
❋ A Brief History of The Earth - Page 10
❋ A Sustainable Way of Thinking - Page 12
❋ Editor's Preface - Page 14
❋ Part I - The Roots of The Environmental Movement - Page 17
❋ Part II - Threats to the Environment - Page 39
❋ Part III - Solutions To The Environmental Crisis - Page 93
❋ Contact List - 50 Ways To Get Involved - Page 136
❋ Suggestions For Further Reading - Page 142
❋ The Teacher's Guide To The Earth Pledge - Page 143
❋ Earth Pledge Essay and Illustration Contest - Page 146
❋ The Urgency of the Earth Pledge - Page 147
❋ The Earth Pledge Song Score - Acoustic Version - Page 149
❋ Ethical Principles To Guide The Earth Pledge - Page 150
❋ Earth Pledge Declaration Form - Page 160

The Earth Pledge gives me hope. It lets me know that there are people everywhere who share the belief that we have been put on this planet as caretakers. It challenges me to make a difference.
Anya Green-Odlum, 8th Grade

❉❉❉

The Earth Pledge is inspiring.
It helps us to remember to protect the earth.
When I hear it it makes me want to just hug someone.
Mary Miller, 5th Grade

❉❉❉

The Earth Pledge is short and simple.
It's message is powerful and timely.
Julia Rinne, Elementary School Principal

The Earth Pledge

I pledge to protect the earth,
And to respect the web of life upon it,
And to honor the dignity
Of every member
Of our global family.
One planet, one people, one world,
in harmony.
With peace, justice and freedom for all.

*****

Objectives of The Earth Pledge Campaign

1. To create a community of hope of people committed to a shared vision of a better, cleaner, safer and healthier natural environment.

2. To provide citizens, especially young people, with an opportunity to make a personal commitment to protect and preserve our natural environment for the use and enjoyment of future generations by becoming "Earth Stewards".

3. To build broad consensus behind a very simple but powerful statement of principles and thereby develop a sense of common purpose.

4. To help create a spirit of cooperation in which each of us is willing to merge our individual talents and agendas into a seamless web of trust and collaboration.

5. To help educate the public about the consequences of our environmental choices.

6. To motivate people to act to incorporate sustainable principles into their daily living habits.

7. To encourage discussion about the power of our words and the importance of our keeping of promises.

8. To provide a user friendly instrument that can help engage the silent majority in the cause of environmentalism.

9. To build bridges of trust and cooperation between schools, universities, businesses, the religious community, the media, the arts and the entertainment community.

10. To have a lot of fun while doing something we believe in.

The Board of Advisors of The Earth Pledge Program

(Partial List - Organizations Listed For Identification Purposes Only)

- Carl Anthony - President, Earth Island Institute, Urban Habitat Program
- Bruce Anderson - President, Earth Day USA
- John Anderson - President, World Federalist Association
- David Brower, Founder - Friends of the Earth, Earth Island Institute
- Dr. Noel J. Brown - Director, United Nations Environment Programme
- Susan Cadogan - President, Earth Angel Inc.
- Fritjof Capra - Founder - President, The Elmwood Institute
- Michael Closson - Executive Director, Center for Economic Conversion
- Jayni Chase - Founder and CEO, Center for Environmental Education
- Mark Cherrington - Editor, *Earthwatch Magazine*
- Paul Coleman - Earth Ambassador Who Walked 8000 Miles To The Earth Summit in Rio
- Don Conroy - President, North American Coalition For Religion and Ecology
- Ram Dass (Richard Alpert) - Founder, Seva Foundation
- Anne and Paul Ehrlich - Center for Population Studies, Stanford University
- Terence Flynn - Chairman of the Board, Portal Publications
- Roy A. Gamse - President, Earth Force
- Lucile Green - Co-Founder and First President, Association of World Citizens
- Paul Hawken - Businessman/Author, "The Ecology of Commerce"

- Gary Herbertson - President, Earth Day International
- Jack Howell - Publisher, Morning Sun Press
- Douglas K. Huneke - Pastor, Westminister Presbyterian Church
- Huey Johnson - President, Resource Renewal Institute
- Joanna Macy - Author, Educator
- Avon Mattison - Founder, Pathways To Peace
- Dr. Rashmi Mayur - President, Global Futures Network
- Michael McCoy - Executive Director, U.S. Citizens Network
- Walter McGuire - President, McGuire and Company
- Victoria Morgan - Publisher, Foghorn Press
- Dr. Robert Muller - Chancellor, University for Peace
- Benton Musslewhite - President, One World Now
- Greg Nepstad - Director, Des Moines Area Greens
- Michele Perrault - International Vice President, The Sierra Club
- Terry Pezzi - President, Terra Christa Communications
- Sandra Postel - Senior Fellow, Worldwatch Institute
- Franko Richmond and Gail Lima - Earth Emerge and See
- Martin Rosen - President, Trust for Public Land
- Armin Rosencranz - President, Pacific Environment and Resources Center
- James W. Rouse - Founder / Chairman, The Enterprise Foundation
- Reverend William Swing - Bishop, California Episcopal Diocese

A Brief History of The Earth

This is a story often told by the environmentalist, David Brower. Perhaps you've heard it. Actually it's not a story. It's fact. It's history. It's a brief history of the Planet Earth.

If we took the history of the planet earth, and condensed it into just one week's time, beginning Sunday morning at midnight, consider these milestones:

☞ Life would not have appeared on the planet until mid-day on Tuesday.

☞ The great dinosaurs appeared on the earth at 4 P.M. on Saturday. They were offstage by 9 that evening.

☞ Homo-sapiens made their entrance at 30 seconds before midnight on Saturday.

•☞ Agriculture began at 1 and 1/2 seconds before midnight.

☞ Jesus Christ made his entrance at 1/4 of a second before midnight.

•☞ The industrial revolution took place at 1/40th of a second before midnight.

☞ World War I took place at 1/100th of a second before midnight.

Now consider what has happened in that last 1/100th of a second:

☞ The population of the Earth has increased threefold.

☞ The population of California has increased by a factor of 12.

☞ Humanity has used 4 times the amount of the earth's resources as were used in all previous history.

How long can we continue this pace of growth and still save our earth for future generations?

It is now midnight ... time to rethink!

A Sustainable Way of Thinking *

Creating a sustainable future will require a different way of thinking. It is not necessarily a new way of thinking. In fact it draws us back to the best and most honored ways of thinking throughout crisis periods in history.

❋ It is a disciplined way of thinking.

❋ It proceeds from a clear sense of purpose.

❋ It is focused on imagining the future in its highest and best form.

❋ It seeks to balance the long-term with the short-term.

❋ It considers the whole web of life in personality, character and spirit.

❋ It seeks ways of shaping institutions to create a sense of community.

❋ It believes that people can live together harmoniously.

❋ It believes that our role is to consciously create a future with promise.

❋ It believes that we must begin to define the common ground as a basis of cooperation, collaboration, trust and understanding.

❋ It believes that meeting these challenges is vital to our survival.

❊ It believes that those engaging in them are at work in the most important task that can possibly consume their lives.

❊ It identifies our primary challenge as finding effective vehicles for engaging diverse sectors of our society including the business community, the religious community, the media, government, universities, secondary schools, non-profits, and the labor community.

❊ It starts from the common assumption that all institutions, no matter how large are composed ultimately of human beings.

❊ It assumes that every human being has a vital stake in the future.

❊ It assumes that it is the fundamental desire of every living person not only to live in a healthy natural environment, but also to participate in the process of creating that environment.

❊ It knows that the very act of creating a workable blueprint for cooperation towards our common goal of sustainability will generate sufficient interest, enthusiasm and tangible resources to invigorate and insure the success of the entire process.

John F. Ince * Inspired by *Way of Thinking* by James Rouse

Editor's Preface

The essence of the Earth Pledge is upon appreciating the power of our words. It is our committed speaking that creates the future. The Earth Pledge is a bold and beautiful vision that holds out the best that is within us as both real and achievable. The Earth Pledge has the power to educate, to motivate, and to inspire us. It is an instrument of commitment and transformation that provides what is called for by the urgency of our environmental crisis: an opportunity to speak into existence sustainable living patterns.

Through our speaking of the Earth Pledge we can begin to create a future with promise by affirming our collective will to truly make a difference. When shared and spoken by enough people it will begin to create transformation of our consciousness. This transformation will create hope, vision and awakening. Please join us in speaking the most important message of our time: the message of environmental awareness and global cooperation.

Environmentalist, David Brower, says, "If you want to save the earth, you must do three things:

- **First you must get people to fall in love with it.**
- **Second, you must explain the threats.**
- **Third, you must tell people what they can do.**

Accordingly, *The Earth Pledge Book* is divided into three parts.

Part I takes us back to the roots of the environmental movement showing us a love of the earth and appreciation of nature.

Part II describes the threats to our future, both in terms of our lifestyles and our ways of relating to the natural world.

Part III offers solutions to the environmental crisis with thoughts that inspire us to action.

Beneath each quotation there is a suggested pledge or commitment. Commitment is what the Earth Pledge is all about. Everything of import or significance begins with a firm commitment. The nature of commitment was perhaps best expressed by the Scottish mountain climber, T. H. Murray when he wrote,

"Concerning all acts of initiative and creation, there is one elementary truth, the ignorance of which kills countless plans and splendid ideas. Until one is committed there is hesitancy, the chance to turn back and always ineffectiveness. But at the moment one commits oneself, then Providence moves too, raising in one's favor all manner of unforeseen incidents meetings and material assistance that no man (or woman) could have dreamed would come their way. I have learned a deep respect for one of Goethe's couplets, 'Whatever you can do or dream, you can at least begin it. For boldness has genius, power and magic in it.' "

• **John F. Ince**

Part I - Roots Of The Environmental Movement

If you were the Mother Earth today, what would you do to save yourself? Why, of course, you would do exactly what she is doing: expressing in mysterious ways, her distress, through us, and through our pain and our anxiety. What is happening all around us is simply Mother Earth reminding us that she is hurting. She is telling us that we must start to restore the balances. She is telling us that we need to rescue her.

How do we rescue our Mother? It all starts with a Pledge. We instinctively protect what we love. Thus, the process of Earth CPR (conservation, protection and restoration) starts with a hike in the hills, a walk through the park or a jog on the beach. It is ultimately the press of our feet to the earth, and our connection with the natural world that helps us replenish our love for the earth. From that love flows a sense of purpose, and a feeling of community and a firm commitment. So the thoughts in Part I have been selected to reacquaint us with the beauty of nature in hopes of inspiring a commitment.   • John F. Ince

Have you listened to the Earth?

Yes the Earth speaks, but only to those who can hear with their hearts.
It speaks in a thousand small ways, but like our lovers and families and friends,
it often sends its messages without words.
For you see, the earth speaks in the language of love.
Its voice is the shape of a new leaf, the feel of a water worn stone, the color of
the evening sky, the smell of summer rain, the sound of the night wind.
The earth whispers are everywhere,
but only those who have slept with it can respond readily to its call. (3)

Steve Van Matre

Pledge Number 1:

I pledge to appreciate the sights, sounds, and smells of nature.

All things are bound together. All things connect.
What happens to the earth happens to the children of the earth.
Man has not woven the web of life. He is but one thread.
Whatever he does to the web, he does to himself.

Chief   Seattle

☆☆☆☆☆

Pledge Number 2

*I pledge to respect all life forms
including animals, plants, insects, and people
of all races, creeds, ethnic backgrounds and nationalities.*

When we saw a few years ago those first
pictures of of the earth from space,
we had a glimpse of... stunning beauty,
that dappled white and blue sphere
stirred us all. (1)

James E. Lovelock

☆☆☆☆☆

Pledge Number 3

I pledge to look upon the earth as one unit.

Parents should be able to comfort their children by saying
"everything's going to be all right", "we're doing the best we can" and
"it's not the end of the world".
But I don't think you can say that to us anymore.
Are we on your list of priorities?
My dad always says "You are what you do, not what you say."
Well, what you do makes me cry at night.
You grown ups say you love us. I challenge you,
please make your actions reflect your words.

Severn Suzuki (12 years old)

☆☆☆☆☆

Pledge Number 3

*I pledge to let my actions do my speaking for me
in helping to save the earth for the use and enjoyment
of future generations.*

Every organism–from the smallest bacterium
through the wide range of plants and animals
to humans–is an integrated whole
and thus a living system.
The same characteristics of wholeness
are exhibited by social systems–such as family or community–
and by ecosystems that consist of a variety of organisms
and inanimate matter in mutual interaction.

Fritjof Capra

Pledge Number 4

*I pledge to respect and honor the rightful place
of every organism in the web of life.
(Contact Number 1)*

The cycles of Life need to be approached with reverence.
They have been in place for billions of years.
They are the reflection of the natural breathing
of the soul of Gaia itself, the earth consciousness,
as it moves its force fields and guides the cycles of Life.
If these are revered, how could we look at something
as exquisite as our Earth's ecology
and do one thing that would risk
the balance of this system?

Gary Zukav

☆☆☆☆☆

Pledge Number 5

I pledge to breathe nature's beauty in deeply.

A child represents the most advanced stage of human evolution.

Theodore Roszak

☆☆☆☆☆

Pledge Number 6

*I pledge to respect the special wisdom of children
and to honor what is in their hearts.*

There is something infinitely healing
in the repeated refrains of nature –
the assurance that dawn comes after night,
and spring after winter.

Rachel Carson

Pledge Number 7

*I pledge to regularly take some time out from my busy routine
to watch a sunrise or a sunset from some peaceful setting.*

In nature you can hear the beat of your own heart
and the sound of your own footsteps.
Take yourself there as often as you can –
to the sea, to the mountains, to the forest and meadows.
In nature, you can learn that
you are part of a grand and beautiful scheme
and that you are as valid a part of life's masterpiece
as is everything that you see around you.

Marilyn Diamond

Pledge Number 8

*I pledge to go for a hike in the hills, a walk through the forest,
a stroll on the beach or visit a nearby park soon.
(Contact Number 2)*

The Earth laughs ... in flowers. (3)

Ralph Waldo Emerson

Pledge Number 9

I pledge t o use organic (chemical free)
lawn and garden care products.
(Contact Number 3)

Up there you go around every hour and a half, time after time...
You look down there and you can't imagine how many borders and
boundaries you cross, again and again and again,
and you don't ever see them. ...
It is so small and fragile and such a precious little spot in the universe
that you can block it out with your thumb,
and you realize that on that little blue and white thing,
is everything that means anything to you–all of history and
music and poetry and art and death and birth and love, tears, joy, games ,
all of it on that little spot out there that you can block out with your thumb.

Russell Schweickart (Apollo 9 Astronaut)

Pledge Number 10

I pledge to think beyond borders and boundaries.

Before 1961 no human being had actually seen
the whole earth, what Plato had described
as "a living creature, one and visible,
containing within itself all living creatures."
It took more than 2,000 years for modern science to rediscover
the ancient wisdom that planet Earth
- the Greeks called it Gaia or Mother Earth -
functions as a single living organism
and that we, as part of the planet, are life within life. (1)

Jon Naar

Pledge Number 11

*I pledge to respect the Gaia principle: that the
earth itself is a living breathing organism.*

When despair for the world grows in me
and I wake in the night at the least sound
in fear of what my life and my children's lives may be,
I go and lie down where the wood drake
rests in his beauty on the water, and the great heron feeds.
I come into the peace of wild things
who do not tax their lives with forethought of grief.
I come into the presence of still water.
And I feel above me the day-blind stars waiting with their light.
For a time I rest in the grace of the world, and am free.

Wendell Berry
☆☆☆☆☆

Pledge Number 12

I pledge to simplify my life.

Deep in their roots,
All flowers keep the light.

Theodore Roethke

☆☆☆☆☆

Pledge Number 13

*I pledge to tend my garden of
flowers, health, happiness and hope.*

(Contact Number 3)

Natural objects themselves,
even when they make no claim
to beauty, excite the feelings
and occupy the imagination.
Nature pleases, attracts delights,
merely because it is nature.
We recognize in it an Infinite Power.

W. Humboldt

☆☆☆☆☆

Pledge Number 14

I pledge to trust the healing power of nature.

All nature is a vast symbolism;
every material fact
has sheathed within it
a spiritual truth.

E.H. Chapin

☆☆☆☆☆

Pledge Number 15

I pledge to honor nature's secrets and respect nature's truths.

Climb the mountains and get their good tidings.
Nature's peace will flow into you
as sunshine flows into trees.
The winds will blow their own freshness into you,
and the storms their energy,
while cares drop off like autumn leaves.

John Muir

☆☆☆☆☆

Pledge Number 16

I pledge to call my local chapter of the Sierra Club and find out about upcoming hikes and expeditions.
(Contact Number 2)

What a joy it is to feel the soft, springy earth
under my feet once more,
to follow grassy roads that lead to ferny brooks
where I can bathe my fingers in a cataract
of ripping notes,
or to clamber over a stone wall into green fields
that tumble and roll and climb in riotous gladness. (3)

Helen Keller

Pledge Number 17

*I pledge to call my local chapter of the Lighthouse For The Blind
and take a blind person for a nature outing.*

The ignorant man marvels at the exceptional;
the wise man marvels at the common;
the greatest wonder of all
is the regularity of nature.

George Dana Boardman

☆☆☆☆☆

Pledge Number 18

I pledge to stop complaining about what doesn't work and appreciate what does, especially nature.

I gave my heart to the mountains the minute
I stood beside this river with its spray in my face
and watched it thunder into foam,
smooth to green glass over sunken rocks,
shatter to foam again... (3)

Wallace Stegner

Pledge Number 19

*I pledge to urge regional planning boards and governing bodies
to adopt green master plans that respect the environment.
(Contact Number 4)*

Surely there is something
in the unruffled calm of nature
that overawes our little anxieties and doubts;
the sight of the deep-blue sky,
and the lustering stars above,
seem to impart a quiet to the mind.

Jonathan Edwards

Pledge Number 20

*I pledge to spend less time inside watching TV and
more time outside in nature getting exercise and fresh air.*

Part II - Threats To The Earth

Something very exciting is happening. And at the same time it is also very scary. We are in the midst of the most profound transformation in human history and it is only beginning to reach our consciousness. We now have our most conclusive evidence of what the human species is doing to the earth. We know about how our explosive population growth is putting stress upon the earth's life support systems. Put simply, it is now dawning on us that we now have the power to inflict unknowable damage to our natural environment through the sum of countless, incremental decisions in the course of daily economic activity. The time has come to accept the challenge posed by the reality of what is happening to the planet earth. There is excitement and empowerment in the accepting of a great challenge. The challenge begins by understanding the outlines of the transformation that is occurring. There are ten foundational precepts of this dawning consciousness:

1) The earth has a carrying capacity.

2) The earth is an organic whole. All its parts including plant life, animal life, the rivers, oceans, mountains, deserts, plains and the human specie all interrelate in an exquisite balance.

3) When the earth's carrying capacity is reached or exceeded, nature's balances are upset and stress is created within the earth's life support systems.

4) The ultimate consequences of upsetting these natural balances cannot be known precisely, but we can say now with a reasonable degree of certainty, that profound ecological shifts are likely.

5) We are now in the midst of the explosive phase of geometrical population growth predicted by Malthus over 100 years ago. We add 1 billion people to the earth every decade. It took 200,000 years for the earth population of the earth to reach 2 billion. In the last 50 years the population of the earth has almost tripled to 5.6 billion.

6) We are now on the verge of exceeding the earth's carrying capacity. In the last 50 years the United States alone has consumed more of the earth's resources than the entire world did in all history.

7) This pace of growth cannot be sustained without an ecological reckoning.

8) No solution to this crisis is possible as we continue to deny and distract ourselves from this reality by our use of drugs, alcohol, cultural escape and all manner of personal and societal violence.

9) To begin to right the balance of the earth, both individuals and institutions must be willing to assume greater responsibility for their actions. Each of us must make a personal Pledge to protect and preserve the earth for future generations. This commitment, made resolutely by a small community will be sufficient to bring about a large scale transformation in society's behavior patterns.

10) This commitment must be made now and must include all sectors of society including business, government, the media, the arts and entertainment community, the educational community, the religious community, the advertising community, the legal community, organized labor, and the non profit world. The sooner we act as a community, the greater our chances of making this transformation as benign as possible.   • John F. Ince

In the last 50 years, the United States alone has consumed more of the earth's resources than the whole world did in all human history.

David Brower

☆☆☆☆☆

Pledge Number 21

I pledge to learn to distinguish between my needs and my wants, and to only consume those things that I really need.
(Contact Number 5)

It took 10,000 generations of human history for
our population to reach 2 billion.
In the last two generations it has gone from 2 billion to 5.6 billion.
If present trends continue, energy consumption
and production of pollutants will double over the next 20 years.

UNESCO

Pledge Number 22

*I pledge to learn the facts about population growth
and the stress it puts on the earth's resources.*

The failure to recognize our unalterable link to the natural world
has resulted in an onslaught of environmental degradation.
The ramifications of our relentless assault on nature are staggering...
The root cause of many of these critical environmental problems
can be traced back to rapid and indiscriminate overdevelopment.
I suggest that the time has come for us to reevaluate our priorities.

Gregory Peck

Pledge Number 23

*I pledge to look carefully at the priorities in my life and
make the environment a top concern.
(Contact Number 6)*

Like most people of my generation,
I was brought up to believe in progress. I still do.
But we are now at a point where we have to ask ourselves if
we are the beneficiaries of our progress–or victims...
Progress from now on has to mean something different...
We're creating a world that is tipping dangerously out of balance.
We are running out of places,
we're running out of resources
and we're running out of time.

Robert Redford

☆☆☆☆☆

Pledge Number 24

*I pledge to consider the effect of "progress"
upon our quality of life.*

It takes no stretch of the imagination to see that the human species
is now an agent of change of geological proportions. ...
At humanity's hand, the earth is undergoing a profound transformation–
one with consequences we cannot fully grasp.
It may be the ultimate irony that in our efforts to make
the earth yield more for ourselves,
we are diminishing its ability to sustain life of all kinds,
humans included.

Sandra Postel
The Worldwatch Institute

☆☆☆☆☆

Pledge Number 25

I pledge to moderate my desires and consume less.
(Contact Number 7)

We believe that the environmental crisis is intrinsically religious.
All faiths, traditions, and teachings
firmly instruct us to revere and care for the natural world.
Yet sacred creation is being violated, and is in the ultimate jeopardy
as a result of long-standing human behavior.
A religious response is essential
to reverse such long-standing patterns of neglect.

Global Forum of Spiritual and Parliamentary Leaders

Pledge Number 26

*I pledge to contact local ministers, rabbis or other religious leaders
and urge them to include concerns of ecology
in prayers and rituals of worship. (Contact Number 8)*

Over the ages of human existence,
the issue of survival arose only in the very beginning:
could the human species evolve through adapting to an often hostile environment...
In a strange reversal of our predicament,
the threat to humanity (today) comes not from a hostile planet
but from the power which man's genius has given him over the planet itself.
It comes from ... our capacity to damage it
and destroy ourselves in the process.

The Stockholm Initiative

Pledge Number 27

*I pledge to urge institutions of higher education to develop
integrative programs of environmental studies.*

By eating low on the food chain you are simply reducing the quantity of most if not all pesticides residues in your diet. If we are wrong and there are no real health hazards ... than no harm has been done. If time shows that accumulated pesticides residues do produce damage to humans, then you may be grateful you heeded this cautionary note.

Francis Moore Lappe

☆☆☆☆☆

Pledge Number 28

I pledge to make more healthful and environmentally sound food choices. (Contact Number 9)

The harsh reality is that no war no revolution,
no peril in all of history measures up
in importance to the threat
of continued environmental deterioration ...
The absence of a pervasive guiding conservation ethic
in our culture is the issue and the problem.
It is a crippling if not indeed fatal weakness.

**Senator Gaylord Nelson,
Founder of Earth Day**

☆☆☆☆☆

Pledge Number 29

*I pledge to make every day Earth Day.
(Contact Number 10)*

Air pollution is turning
Mother Nature prematurely gray.

Irv Kupcinet

☆☆☆☆☆

Pledge Number 30

*I pledge to make sure my car complies with
clean air standards and is equipped with smog control devices.
(Contact Number 11)*

We cannot command nature
except by obeying her.

Francis Bacon

☆☆☆☆☆

Pledge Number 31

*I pledge to listen more attentively to
the message Mother Earth is sending us.*

The natural resources of this planet,
including its human resources,
are the ultimate source of all economic activity.
Exhaust the soil, poison the water,
change the balance of the atmosphere,
and you curtail our own future.

**John H. Adams,
The Natural Resources Defense Council**

Pledge Number 32

*I pledge to become better informed about the dangers of
resource depletion and water pollution.
(Contact Number 12)*

The American earth represents our greatest heritage.
The right to pure water, streams, virgin forests, the woodchuck
and the antelope, and the other exciting wonders of the woods
are as basic to our freedom as the special rights
enshrined in our bill of rights.

Justice William O. Douglas

Pledge Number 33

*I pledge to learn how I can provide legal
protection to our natural wonders.
(Contact Number 13)*

The dinosaur's eloquent lesson
is that if some bigness is good,
an overabundance of bigness
is not necessarily better.

Eric Johnson

☆☆☆☆☆

Pledge Number 34

*I pledge to learn the value
of moderation.*

The mastery of nature is vainly believed
to be an adequate substitute for self-mastery.

Reinhold Niebuhr

Pledge Number 35

*I pledge to conquer my own fears
before I try to conquer the world.*

Nature thrives on patience.
Man on impatience.

Paul Boese

Pledge Number 36

I pledge to live my life at a more natural pace.

The United States consumes more energy
for air conditioning
than the total energy consumption
of the one billion people in China.

Robert O. Anderson

Pledge Number 37

*I pledge whenever possible to open the windows
to let fresh air in rather than use an air conditioner
and to plant trees for shade, which can reduce
air conditioner needs by 50%.*

Faced with widespread destruction of the environment,
people everywhere are coming to understand that
we cannot continue to use the goods of the earth
as we have in the past.
A new ecological awareness is beginning to emerge,
which rather than being downplayed, ought to be encouraged
and developed into concrete programs and initiatives.

Pope John II

☆☆☆☆☆

Pledge Number 38

*I pledge to work with others in my church or synagogue
to develop a local environmental initiative or program.*

Pollution along our coast and beaches
appears to have become the rule rather than the exception.
The International Coastal Cleanup, the world's largest,
collected 3.7 million pounds of trash along 4,743 miles of beach.
The cleanup enlisted more than 145,000 volunteers
in 43 states and territories and 12 countries.
Plastics were the most common debris item reported,
accounting for 59 percent of all debris collected.

Thomas Miller

Pledge Number 39

*I pledge to put my beach trash in receptacles and especially
to make sure plastic sixpack holders are disposed of properly.*

Rapidly expanding population effectively
strangles most efforts to provide
adequate education, nutrition, health care
and shelter.

Gro Harlem Brundtland

Pledge Number 40

*I pledge to learn to distinguish between
symptoms of problems
and fundamental causes of problems.*

Such prosperity as we have known it
up to the present is the consequence
of rapidly spending the planet's irreplaceable capital.

Aldous Huxley

☆☆☆☆☆

Pledge Number 41

*I pledge to conserve energy and
use renewable resources wherever possible.
(Contact Numbers 23 and 24)*

A citizen of an advanced industrialized nation
consumes in six months the energy and raw materials
that have to last the citizen
of a developing country his entire lifetime.

Maurice Strong

Pledge Number 42

*I pledge to reduce energy consumption in transportation
by taking the bus, carpooling, riding a bicycle or walking
instead of driving my car whenever possible.*

What is the use of a house
if you haven't got a tolerable planet to put it on?

Henry David Thoreau

☆☆☆☆☆

Pledge Number 43

*I pledge to take the long term view
when making important decisions.*

A leading local environmentalist...asked me if I was actually optimistic about the future.
I told him that, in fact, I am optimistic, but I must admit I was puzzled
why there should be such a difference between us. ...
The answer, I came to see, is a function of our own environments.
He spends all his time working with politicians and activists,
and faces a daily sense of frustration.
I spend my time dealing with volunteers and scientists,
so I have a daily diet of accomplishment.

Mark Cherrington, Earthwatch

Pledge Number 44

*I pledge to become actively involved in solving environmental problems
rather than just complaining about them.
(Contact Number 14)*

Human activities inflict harsh and often irreversible damage
on the environment on critical resources.
If not checked, many of our current practices
put at serious risk the future that we wish for human society
and the plant and animal kingdoms, and may so alter the living world
that it will be unable to sustain life in the manner that we know.
Fundamental changes are urgent if we are
to avoid the collision our present course will bring about.

The Union of Concerned Scientists
(Signed by over 1700 eminent scientists)

☆☆☆☆☆

Pledge Number 45

I pledge to use emerging technologies that
are more friendly to the environment. (Contact Number 15)

In the past, Earth's bounty appeared limitless;
nothing people could do, it seemed, would ever deplete that bounty.
Today, of course, we know that we cannot
systemically exploit our natural resources without adverse consequences.
We know that we cannot pollute our rivers,
fill in our wetlands and indiscriminately dump toxins
without affecting public health. ...
Clearly, the long-term health of people depends on the
health of our global life-support system–the environment.

John C. Sawhill, President, The Nature Conservancy

Pledge Number 46

I pledge to work to preserve our natural habitats.
(Contact Number 16)

Planting a tree is a spiritual as well as a physical act–
helping set forth a new life in a new place–
a life that could with luck, run hundreds of years
and affect many generations of people.

R. Neil Sampson

Pledge Number 47

I pledge to plant a tree in my community soon.

The more we exploit nature,
the more our options are reduced,
until we have only one;
to fight for survival.

Morris Udall

Pledge Number 48

*I pledge to learn to live in harmony with nature
rather than exploiting nature.
(Contact Number 17)*

A nation without the means of reform
is without means of survival.

Edmund Burke

☆☆☆☆☆

Pledge Number 49

*I pledge to write members of congress and
urge them to put the future of our planet
ahead of their own political futures.
(Contact Number 18)*

Thy command, "Be fruitful and multiply."
was promulgated, according to our authorities,
when the population of the world
consisted of two persons.

Dean William R Inge

☆☆☆☆☆

Pledge Number 50

I pledge to support family planning worldwide.
(Contact Number 19)

Overpopulation occurs
when people take leave of their census.

Malcolm K Jeffrey

Pledge Number 51

I pledge to support Third World education programs aimed at empowering women in making decisions about their reproductive rights.

We all worry about the population explosion –
but we don't worry about it at the right time.

Art Hoppe

☆☆☆☆☆

Pledge Number 52

*I pledge to think before I bring
another human being in the world.*

Man thinks of himself as a creator instead of a user,
and this delusion is robbing him,
not only of his natural heritage,
but perhaps of his future.

Helen Hoover

☆☆☆☆☆

Pledge Number 53

*I pledge to use more of my time creating resources
and less time consuming the earth's resources.*

We abuse land because we regard it
as a commodity belonging to us.
When we see land as a community
to which we belong,
we may begin to use it
with love and respect. (3)

Aldo Leopold

☆☆☆☆☆

Pledge Number 54

*I pledge to join together with others in my school,
business, family or community
to conserve, protect and restore the land.
(Contact Numbers 20 and 21)*

I durst not laugh,
for fear of opening my lips
and receiving the bad air.

Shakespeare

☆☆☆☆☆

Pledge Number 55

*I pledge to do what I can to help reduce
air pollution in my community.
(Contact Number 22)*

Up to 75% of the electricity produced in the U.S.
is wasted through the use
of inefficient motors, lights and appliances. (1)

David Moskowitz

☆☆☆☆☆

Pledge Number 56

*I pledge to conserve and use energy
more efficiently in my home, school or place of work.
(Contact Numbers 23 and 24)*

The goal of the Cold War
was to get others to change
their values and behavior,
but winning the battle to save the planet
depends on changing our own values and behavior.

Lester R. Brown
Worldwatch Institute

Pledge Number 57

I pledge to change my own behavior before
I try to get everyone else to change theirs.
(Contact Number 25)

The supreme reality of our time is ...
the vulnerability of our planet.

John F. Kennedy

Pledge Number 58

*I pledge to tell four friends about the
Earth Pledge Campaign and ask them to write the
President urging his support for national
and international programs of Earth CPR
(Conservation, Protection and Restoration).
(Contact Number 26)*

We have come to a juncture
in humanity's evolution,
where we must call a halt
to our uncontrolled assault
on the environment. (4)

Jimmy Carter

☆☆☆☆☆

Pledge Number 59

I pledge to write the major newspapers and news networks and urge them to give more coverage to environmental issues.

It is good business to anticipate the inevitable, and it seems to me inevitable,
whether we like it or not, that we are moving toward an economy which
must be limited and selective in its growth pattern.
The earth has finite limits–a difficult idea for Americans to adjust to ...
Yet the realization is coming home to us with increasing force
that our planet will hold only a limited number of people,
that we must limit the way in which we use its resources,
that our economic system must therefore have its basic
limitations as well.

John D. Rockefeller III

☆☆☆☆☆
Pledge Number 60

*I pledge to urge international financial institutions
such as IMF and The World Bank to integrate
sustainability into lending decisions. (Contact 27)*

We can expect that as the world population increases,
each person's share of the earth's resources
will dwindle despite all that science can do.
It is true that there are vast reserves
of such resources as fossil fuels,
but as time goes by these supplies will be of lower grade
and progressively more difficult to extract.

Stewart   L.   Udall

Pledge Number 61

*I pledge to urge my local school, church or university
to include a learning module about population issues.*

Nature is the most thrifty thing in the world;
she never wastes anything; she undergoes change,
but there's no annihilation–
the essence remains.

T. Binney

☆☆☆☆☆

Pledge Number 62

*I pledge to use cloth napkins, diapers and towels
instead of using disposable ones.*

Each time some close-to-home field or woodlot
is slated for the bulldozer–
each time some swamp is to be drained,
some hilltop leveled, some stream channeled,
some meadow paved–we must weigh carefully
not only the anticipated biological losses
but the loss to our own soul as people.

Martin J. Rosen
Trust For Public Land

Pledge Number 63

I pledge to get involved in community conservation projects.
(Contact Number 28)

You simply may not have a continually expanding economy
within a finite system: Earth.
At least not if the economy is based upon anything approaching
the technological exploitation and production as we now know it.
On a round ball, there is only so much of anything,
Minerals, Food, Air, Water, Space
... and things they need to stay in balance.

Jerry Mander

☆☆☆☆☆

Pledge Number 64

*I pledge to write my elected representatives and urge them
to support incentives for solar and renewable energy.
(Contact Number 29)*

Acid rain spares nothing.
What has taken humankind decades to build
and nature millennia to evolve is being impoverished and
destroyed in a matter of a few years –
a mere blink in geologic time. (1)

Don Hinrichsen

☆☆☆☆☆

Pledge Number 65

*I pledge to do a home energy audit
that includes checking walls and floors for leaks,
closing doors to rooms seldom used and turning off the
heating or air conditioning.*

The disappearance of tropical forests,
home to over half of all species of life on Earth,
is perhaps the most significant ecological disaster we face.
An area of tropical forest
at least the size of Washington state is being destroyed each year.
Tropical deforestation is now dooming 50,000 species
of plants and animals to extinction every year.
This frightening genetic erosion includes a loss
of important resources for foods, medicines and industrial products.

Greenpeace

☆☆☆☆☆

Pledge Number 66

*I pledge to learn what I can do to help prevent the
destruction of the rainforests. (Contact Numbers 17 and 30)*

But burgeoning urban populations and farm use are draining
water supplies, thus increasing the value of water.
All across the continent, water availability is becoming an issue.....
A new generation of water pollutants has created the second major issue.
Toxic materials threaten water supplies once considered pure
if they contained no disease-carrying bacteria or viruses.
But the U.S. Environmental Protection Agency now considers
toxic pollution in drinking water supplies
to be one of the greatest environmental hazards in America.

World Resources Institute
☆☆☆☆☆

Pledge Number 67

*I pledge to contact the World Resources Institute and find
out what I can do to influence and help protect our water supplies.
(Contact Number 31)*

I couldn't wait for our future to get here.
I was sure we would all share
in the glories of better living through technology,
bringing our pursuit of happiness to a triumphant end.
It all made so much sense! So what happened?
What happened to our future?...
The dynamic of that idea swings upon a vicious counter-balance...
Our future wasn't supposed to be like this. (4)

Tom Hanks

Pledge Number 68

*I pledge to write the major movie studios and urge them
to incorporate environmental themes into more movies.
(Contact Number 32)*

We have reached an unsettling and portentous
turning point in industrial civilization.
Because corporations are the dominant institutions
on the planet, they must squarely address
the social and environmental problems that afflict humankind.
To create an enduring society we will need
a system of commerce and production where each and every act is
inherently sustainable and restorative.

Paul Hawken

☆☆☆☆☆

Pledge Number 79

*I pledge to write local businesses urging a comprehensive
environmental audit to assess and mitigate
their adverse impacts on the earth.*

Some scientists now predict a four to nine degree Fahrenheit rise in the
earth's surface temperature over the next fifty years as we continue
to spew carbon dioxide, methane, chlorofluorocarbons, and nitrous oxides
into the atmosphere blocking solar heat from escaping the planet....
The earth's temperature has not varied more than about
four degrees Fahrenheit since the last ice age 18,000 years ago.
Now scientists project a change in temperature in less than one generation
that may well exceed an entire geological epoch in world history.

Jeremy Rifkin

Pledge Number 70

*I pledge to reduce my use of fossil fuels by
keeping my thermostat down and
by insulating and weatherstripping my home.*

For every 1% decrease in ozone,
experts believe that there may be a 2% increase
in ultraviolet radiation (UV-B) and,
according to the U.S. Environmental Protection Agency (EPA),
an estimated 20,000 additional deaths yearly
from related skin cancer.

Geoffrey C. Saign

☆☆☆☆☆

Pledge Number 71

I pledge to reduce the amount of CFC's I produce by using my car air conditioner only when absolutely necessary and making sure that CFC's are recycled when my air conditioner is repaired.

Part III - Solutions and Hopes For The Future

In Part III you will find others reaching out to you and offering their knowledge, solutions, and hopes for the future. It is my hope that in reading this section you will draw sufficient strength to proceed in a purposeful, informed way. It is my hope that this part will help you to emerge from your isolation, and to reach out to others in your communities, your businesses, your families, and your schools, to co-create some finite achievable project of Earth Stewardship and Earth CPR: conservation, protection, and restoration. Through our collective commitment we will begin the process of consciously creating a future with promise. In so doing we will also create hope in our own lives.

If you ever start to despair as you begin the process of engagement, just hold this one thought in your consciousness. Mother Earth is finding her own ways of communicating a fundamental truth through us. Her message is communicated in the way we are feeling and behaving as individuals and as a society.

Yes, we sometimes feel alone in our despair. Sometimes we feel we cannot go on. But Mother Earth, is telling us exactly what we need to know in order to heal ourselves. She is telling us that we can't continue in our present mode of thinking. She is bringing us full circle in our consciousness, back towards an awareness of the earth itself as a living, breathing, growing organism. In this way we begin to realize that the earth is much more than a source of mineral resources provided for our consumption.

We live by abstraction. We read into others what we feel about ourselves. We believe that if we can find the energy, and resourcefulness within ourselves to move ahead purposefully, then others can do the same. This is stuff of real progress: a progress founded upon a realistic assessment of the limits imposed by nature. So read on to get acquainted with others in our community of hope for the future and to make your Pledge to help save the earth.

• **John F. Ince**

People are not going to find their truth-force
in listening to the experts, but in listening to themselves,
for everyone in their way is an expert on
what it is to live on an endangered planet.

Joanna Macy

☆☆☆☆☆

Pledge Number 72

*I pledge to listen carefully to arguments on both sides
of environmental issues, before I make up my mind.*

I have come to believe that we must take bold and unequivocal action:
we must make the rescue of the environment
the central organizing principle for civilization.
Whether we realize it or not, we are now engaged in an epic battle to
right the balance of our earth, and the tide of this battle will turn
only when the majority of people in the world become sufficiently aroused
by a shared sense of urgent danger to join an all-out effort.
It is time to come to terms with exactly how this can be accomplished. (2)

Al Gore

☆☆☆☆☆

Pledge Number 73

*I pledge to become aroused by the urgency of our
environmental crisis and actively involved in Earth CPR.
(Contact Number 34)*

Never doubt that a small group
of committed citizens can change the world.
Indeed it is that only thing that ever has.

Margaret Mead

☆☆☆☆☆

Pledge Number 74

*I pledge to act to the best of my ability
to make the earth a secure and hospitable home
for present and future generations.**

* From Earth Summit Pledge by Maurice Strong

It is high time for humanity to accept and work out
the full consequences of
the total global and interdependent nature
of our planetary home and of our species.
Our survival and further progress will depend largely on the advent
of global visions and of proper global education
in all countries of the world.

Dr. Robert Muller

☆☆☆☆☆

Pledge Number 75

*I pledge to write the United Nations and urge them
to revise their charter to make it more responsive
to the modern environmental challenge.
(Contact Number 35)*

Environmental solutions
that make economic sense
are needed now more than ever.

**Fred Krupp
Environmental Defense Fund**

☆☆☆☆☆

Pledge Number 76

*I pledge to practice a new eco-nomics when
managing personal and business finances.
(Contact Number 36)*

There is a part of every human being that yearns to be whole-
connected to every life form that is,
from the smallest cell to the most complex creatures.
When we know and live and teach our connection to all that is,
then we shall know wholeness.

Foundation for Global Community

Pledge Number 77

I pledge to strive towards wholeness
of mind, body and spirit.
(Contact Number 37)

Human history
becomes more and more a race
between education and catastrophe.

H. G. Wells

☆☆☆☆☆

Pledge Number 78

*I pledge to contact my local school board and school
principal and urge them to explore innovative ways to
integrate issues of sustainability with all studies.
(Contact Number 38)*

As of 1992 there were almost 4000 curbside recycling programs
in the United States–an increase of over 250% since 1988.
Much attention has been given to the costs of curbside
recycling programs, but few Americans are aware that
a dozen US communities are recovering 50 percent
and more of their municipal solid waste
at costs far below traditional collection and disposal methods.

Earth Journal

☆☆☆☆☆

Pledge Number 69

*I pledge to reduce, reuse and recycle.
(Contact Number 33)*

Sustainable development means
development that meets the needs of the present
without compromising
the ability of future generations
to meet their own needs.

Our Common Future

☆☆☆☆☆

Pledge Number 80

*I pledge to honor the interests of future generations
in my present choices.*

The salvation of this human world
lies nowhere else than in the human heart...
in the human power to reflect ...
in human meekness...
and in human responsibility.

Vaclav Havel

☆☆☆☆☆

Pledge Number 81

*I pledge to spend some quiet time every day
reflecting upon what really matters.*

If we fail to seize the moment,
history will never forgive us -
if there is a history.

Thomas A. Watson

Pledge Number 82

*I pledge to learn more about history and
to work actively towards creating a future with promise.*

If we can begin to heal ourselves,
then I believe that process
can be applied to our own society
and even the world at large.

Eileen Rockefeller Growald

☆☆☆☆☆

Pledge Number 83

I pledge to think globally and act locally.

A phrase briefly fashionable in the mid-20th century–
"citizen of the world" will have assumed real meaning
by the end of the 21st century.
Collective action on a global scale will be easier
to achieve in a world already knit together
by cables and airwaves.

Strobe Talbott

Pledge Number 84

*I pledge to look beyond race, religion, ethnic origin, or nationality
and to accept everyone as a fellow "citizen of the world".
(Contact Number 39)*

We all know that the air we breathe and water we drink
needs to be cleaner. We all know that the power we use and
the items we manufacture are depleting our natural resources.
And we all know that now is the time to take action.
Although there may be debate over the particulars,
there is broad general agreement that today's smart business
must keep an eye on both the bottom line,
and a responsibility to the environment.

Christian W.E. Haub, President
The Great Atlantic and Pacific Tea Company

Pledge Number 85

*I pledge to incorporate environmental considerations
in calculations of the bottom line.*

The world is green and beautiful
and God has appointed you
His stewards over it. (2)

Mohammed

Pledge Number 86

*I pledge to become more aware of my personal responsibility
for making the earth cleaner, healthier and more
beautiful.
(Contact Number 40)*

Earth teaches us patience, love;
Air teaches us mobility, liberty;
Fire teaches us warmth, courage;
Sky teaches us equality, broad-mindedness;
Water teaches us purity, cleanliness. (2)

Guru Granth Sahib

Pledge Number 87

*I pledge to be more patient, loving, courageous,
broadminded and clean living.*

The most effective tool we have as citizens, as parents,
is the sheer force of numbers.
When confronted with superior size and strength, dolphins band together.
They attack power with wisdom.
We, too can rise as a human family to stop environmental destruction,
to safeguard the legacy of life we must pass on to future generations.
We can take our inspiration from the dolphins.

Jacques-Yves Cousteau

Pledge Number 88

*I pledge to take my inspiration from the dolphins and join together
with other concerned citizens to stop environmental destruction.
(Contact Number 41)*

We cannot segregate the human heart
from the environment outside us
and say that once one of these is reformed
everything will be improved.
Man is organic with the world.
His inner life molds the environment,
and is itself deeply affected by it.
The one acts upon the other and
every abiding change in the life of man
is the result of these mutual reactions. (2)

Baha'i

Pledge Number 89

I pledge to honor the best feelings within my heart.

The fate of mankind,
as well as of religion
depends upon the emergence of
a new faith in the future. (2)

Teilhard de Chardin

Pledge Number 90

*I pledge to be open to the emergence of
a stronger faith in the future and to the possibility
of miracles in our lives.*

More than anything else we need a new generation of leaders
in the cause of environmentalism.

Jay D. Hair
National Wildlife Federation

☆☆☆☆☆

Pledge Number 90

I pledge to become a leader for the environment
within my school, business or community.
(Contact Number 42)

A world that has become a single geographic unit is now
groping its way, however slowly, towards global institutions
as the only way of achieving common safety and common progress.
A new world is waiting to be born.

Norman Cousins

☆☆☆☆☆

Pledge Number 91

*I pledge to support initiatives and institutions
that facilitate global cooperation
and global solutions to global problems.*

Man shapes himself through decisions
that shape his environment.

Rene Dubos

Pledge Number 92

*I pledge to consciously shape a future with promise
by protecting and preserving a healthy
natural environment in my
home and community.*

A man is ethical only when life,
as such, is sacred to him,
that of plants and animals
as well as that of his fellow man,
and when he devotes himself
helpfully to all life that is in need of help.

Albert Schweitzer

Pledge Number 93

*I pledge to dedicate part of my life
to service of the environment.*

Our ideals, laws and customs
should be based on the proposition
that each generation in turn becomes the custodian
rather than the absolute owner
of our resources-and each generation
has the obligation to pass this inheritance on to the future.

Alden Whitman

☆☆☆☆☆

Pledge Number 94

*I pledge to treat the earth's resources as a sacred inheritance
and to pass them on to my heirs in good order.*

Public concern has at last driven political leaders
to acknowledge what scientists
have been telling us for years -
that river based pollution, the burning of forests,
acid rain, destruction of the ozone layer
and global warming transcend national boundaries
and must have international solutions.

Bruce Babbitt

Pledge Number 95

*I pledge to write world leaders and urge them to
cooperate with each other to solve global environmental problems.
(Contact Number 43)*

We are going to have to find ways of
organizing ourselves cooperatively,
sanely, scientifically, harmonically and in regenerative spontaneity
with the rest of humanity around earth ...
We are not going to be able to
operate our spaceship earth successfully nor for much longer unless
we see it as a whole spaceship and our fate as common.

Buckminster Fuller

☆☆☆☆☆

Pledge Number 96

*I pledge to join or help organize an environmental club
or group in my school, community or place of business.*

Mass communications have a powerful and growing influence
on policy makers in every part of our shrinking globe.
Especially in the United States,
press coverage often determines which issues
get urgent attention from Congress and the White House.
Getting governments to act on critical issues increasingly requires
the mobilization of world opinion through the media.

Population Action International

Pledge Number 97

*I pledge to write letters to the editor of my local newspaper,
radio, and TV Stations urging them to give more attention
to population issues and the environment. (Contact Number 19)*

We need to make a world
in which fewer children are born,
and in which we take better care of them.

George Wald

Pledge Number 98

*I pledge to only have children for whom
I can provide care and love.
(Contact Number 44)*

Even the most casual reading of the earth's vital signs
immediately reveals a planet under stress.
In almost all the natural domains, the earth is under stress–
it is a planet that is in need of intensive care.
Can we pioneer sustainable patterns of consumption
and lifestyle, and can we educate for that?
This is a challenge that I would like to put to you?

Dr. Noel J. Brown

☆☆☆☆☆

Pledge Number 11

*I pledge to urge my children, students, family and friends
to learn what it means to live sustainably
and to lead them by example.*

As consumers,
we have real power to effect change.
We can use our ultimate power,
voting with our feet and our wallets. (1)

Anita Roddick

☆☆☆☆☆

Pledge Number 100

*I pledge to be more conscious of the environmental
implications of my shopping decisions.
(Contact Numbers 6 and 45)*

Nobody made a greater mistake than
he who did nothing because he could only do a little.

Edmund Burke

Pledge Number 101

*I pledge to do something, however small,
to conserve, protect and restore the earth.*

We are awakening to a new moral sense in our religious institutions,
a moral sense that is finding expression in the doctrine
of stewardship of the human in relation to the entire planet.
While this concept of "stewardship" is much needed
it needs to be completed with the doctrine of "communion"
with the natural world. A complete moral consciousness should bring
about a mutual intimacy of the human with
the entire community of living things.

Thomas Berry

☆☆☆☆☆

Pledge Number 102

*I pledge to urge my church or religious group to become
awakened to our moral responsibility to protect the earth.*

If society maintains its current reliance on growth to solve short-term problems,
we believe that population and material production will grow past sustainable limits,
that the carrying capacity of the earth will be eroded,
and that there will then be an uncontrolled decline in population and economic activity.
However this outcome does not appear inevitable.
Mankind could instead begin to assess realistically the limits to material growth.
Society's goals and institutions could be altered to reduce growth now
and more ultimately towards an orderly accomodation
with the finite constraints of the globe.

Donella and Dennis Meadows.

☆☆☆☆☆

Pledge Number 103

*I pledge to realistically assess the limits to growth
and to adjust my life accordingly.*

Environmental regulations have also led to
the creation of a new environmental industry.
It's young now, and still relatively small–
but indications are that it will remain a permanent fixture
on the American scene.
Some well known corporations have
gotten into the field in a substantial way.

Russell Peterson

☆☆☆☆☆

Pledge Number 104

*I pledge to write the presidents of major corporations
urging them to diversify into environmental protection.
(Contact Number 46)*

Let everyone sweep in front of his or her own door
and the whole world will be clean.

Goethe

☆☆☆☆☆

Pledge Number 105

*I pledge to clean up my own mess
and to be responsible for the
consequences of my
own actions.*

By seeing that we are truly one with all things, we see that
in supporting the killing of others or the destruction of the natural world,
we are inviting that destruction upon ourselves.
With this realization we begin the true healing,
which means coming into resonance
with the creator's one law:
You shall be in good relationship with each other
and with all things in the Great Circle of Life.

Brooke Medicine Eagle

☆☆☆☆☆

Pledge Number 106

*I pledge to move beyond adversarial relationships
to cooperation, collaboration and coordination
with all beings in our efforts to save the earth.*

In the long run, it is the sum total
of the actions of millions of individuals
that constitute effective group action ...
get involved in political action.
Otherwise, we shall all eventually
find ourselves stranded in space on a dead Spaceship Earth
with no place to go and no way to get there.

Dr. Paul R. Ehrlich

☆☆☆☆☆

Pledge Number 107

I pledge to get involved in environmental action.
(Contact Number 47)

This is our world, and it's the only one we've got.
We can make these next ten years the
'decade of the environment' or the
"decade of demise."
Which will it be? You can make the difference.

Diane MacEachern

Pledge Number 108

*I pledge to become actively involved in Earth CPR
(Conservation, Protection and Restoration)
by contacting some of the organizations
listed on page 136 of this book .
(Contact Number 48)*

Conservation means development as much as it does protection.
I recognize the right and duty of this generation to develop
and use the natural resources of our land;
but I do not recognize the right to waste them, or to rob,
by wasteful action,
the generation that comes after us.

Theodore Roosevelt

✩✩✩✩✩

Pledge Number 109

I pledge to respect the rights of coming generations.
(Contact Numbers 41 and 49)

Our children...bring with them
the empowerment of their own optimism and
the belief in their own omnipotence that only belongs to the young.
By putting the means at their disposal, and giving them a voice,
we form the strongest coalition imaginable with our children,
the past with the future, knowledge with faith,
a covenant they'll keep with their children
that may help close up the hole in the sky.

Meryl Streep

Pledge Number 110

*I pledge to be more optimistic about the future and
to do what I can to empower children to help the environment.
(Contact Number 50)*

Accuse not Nature,
she hath done her part;
Do thou ... thine!

John Milton

☆☆☆☆☆

Pledge Number 111

I pledge to do my part to help save the earth.

Contact List 50 - Ways To Get Involved

1. The Center For Eco Literacy, 2522 San Pablo Ave, Berkeley, CA, 94702; (510) 845-4595.
2. Sierra Club, 730 Polk Street, San Francisco, CA 94109; (415) 776-2211.
3. National Audubon Society, 833 Third Avenue, New York, NY 10022; (212) 832-3200 or the National Wildflower Research Center, 2600 FM 973 North, Austin Texas, 78725-4201; (512) 929-3600.
4. The Resource Renewal Institute: Fort Mason Center, Building A, San Francisco, CA 94123; (415) 928-3774.
5. The Earth Island Institute: 300 Broadway, Suite 28, San Francisco, CA 94133; (415) 788-3666.
6. The Council on Economic Priorities, 30 Irving Place, New York, NY 10003 For a copy of Shopping for a Better World (Rates 2,400 Brand Items and Corporations) (212) 420-1133.
7. Renew America, 1001 Connecticut Avenue, NW, Washington DC 20036; (202) 232-2252.

8. North American Coalition For Religion and Ecology, 5 Thomas Circle, NW, Washington DC 20005; (202) 462-2591.
9. EarthSave, 706 Frederick Street, Santa Cruz, 95062; (408) 423-4069.
10. Earth Day USA, PO Box 470, Peterborough, NH, 03458; (603) 924-7720.
11. Environmental Protection Agency, 401 M Street SW, Washington DC 20460; (202) 541-4040.
12. Natural Resources Defense Council at 40 West 20th Street, New York, NY 10011; (212) 727-2700 or Friends of the Earth, 218 D Street, SE, Washington DC 20003; (202) 544-2600.
13. World Wildlife Fund, 1250 Twenty-fourth Street, NW, Washington DC 20037; (202) 293-4800 or the National Wildlife Federation, 1400 16th Street., NW, Washington DC 20036; (202) 832-3200.
14. Earthwatch: 680 Mount Auburn St. P.O. Box 403, Watertown, MA 02272-9924; (617) 926-8200, Fax: (617) 926-8532
15. Union of Concerned Scientists, 26 Church Street, Cambridge, MA 02238; (617) 547-5552.
16. The Nature Conservancy, 1815 North Lynn Street, Arlington, VA 22209; (703) 841-5300.
17. Greenpeace, 1436 U Street NW, Washington DC 20009; (202) 462-1177.

18. Representative _____, United States House of Representatives, Washington DC 20510 or Senator _____, United States Senate, Washington DC 20500.
19. Population Action International, 1120 19th Street, N.W. Suite 550, Washington DC 20036-3605; (202) 659-1833.
20. Citizens for a Better Environment, 500 Howard Street #506, San Francisco, CA 94105; (415) 243-8373.
21. National Coalition Against Misuse of Pesticides, 530 Seventh Street, SE Washington DC 20003; (202) 543-5450.
22. National Clean Air Coalition, 530 Seventh Street, SE, Washington DC 20003; (202) 543-8200.
23. Rocky Mountain Institute, 1739 Snowmass Creek Road, Snowmass, CO 81654-9199; (303) 927-3851.
24. Conservation and Renewable Energy Inquiry Service, P.O. Box 8900, Silver Spring, MD 20907; (800) 523-2929. or American Solar Energy Society, 2400 Central Avenue, Unit B-1 Boulder, CO 80301; (303) 443-3130.
25. The Worldwatch Institute, 1776 Massachusetts Ave., NW, Washington DC 20036 (202) 452-1999.
26. The White House, 1600 Pennsylvania Ave, Washington DC 20050.
27. 50 Years is Enough, 1025 Vermont Ave NW # 300, Washington DC 20005

28. Trust For Public Land, 116 New Montgomery Street, 4th Floor, San Francisco, CA 94105; (415) 495-4014.
29. Solar Energy Industries Association, 122 "C" Street, NW, 4th Floor, Washington DC 20001; (202) 408-0660.
30. Rainforest Action Network, 301 Broadway, Suite A, San Francisco, CA 93133; (415) 398-4404.
31. World Resources Institute, 1735 New York Avenue NW, Washington DC 20006; (202) 638-6300 or for publications (800) 822-0504.
32. Environmental Media Association, 3679 Motor Ave, Suite 300, Los Angeles, CA 90034; (310) 287-2830.
33. For a list of state recycling offices call EPA Office of Solid Waste (800) 424-9346 or the Environmental Defense Fund recycling line (800) CALLEDF.
34. Vice-President Gore, The White House, 1600 Pennsylvania Avenue, Washington DC 20050.
35. H. E. Boutros Boutros-Gali, Secretary General of the U.N., 3-3800, United Nations, New York, NY 10017, 212-963-5015 or the Campaign For U.N. Reform, 418 Seventh Street S.E., Washington DC 20003; (212) 546-3956.

36. Environmental Defense Fund, 257 Park Avenue South, New York, NY 10010; (212) 505-2100.

37. Foundation For Global Community, 222 High Street, Palo Alto, CA 94301-1907; (415) 328-7756.

38. North American Association for Environmental Education, Memberships and Publications, P.O. Box 400, Troy, Ohio 45373; (513) 676-2514; Washington DC Headquarters, Suite 400, 1255 23rd St. NW, Washington DC, 20037 or contact the Center For Environmental Education, 881 Alma Real Drive, Suite 30, Pacific Palisades, CA 90272; (310)454-4585.

39. The Campaign For Global Change, The World Federalist Association, 418 7th Street, SE, Washington DC 20003-2796; (800) WFA-0123

40. Pew Global Stewardship Initiative: 1333 New Hampshire Avenue, NW, Suite 1070 Washington DC 20036; (202) 736-5800.

41. The Cousteau Society, 870 Greenbrier Circle, Suite 402, Chesapeake, VA 23320-2641; (804) 523-9335.

42. National Wildlife Federation, 1400 Sixteenth Street, N.W., Washington D.C. 20036-2266; (202) 797-6800.

43. Refer to: *How To Write World Leaders* by Rick Lawlor; Contact: MinRef Press, 8379 Langtree Way, Sacramento, CA 95823; (916) 424-8465.

44. Zero Population Growth, 1400 16th Street NW, Washington, DC 20036. Ask for their "Ease the Squeeze" poster with its Earth-friendly facts and action tips; (202) 332-2200.

45. Co-op America, 2100 M Street NW, Washington DC 20063; (202) 872-5307.

46. CERES-Coalition for Environmentally Responsible Economies, 711 Atlantic Avenue, Boston, MA 02111; (617) 451-0927.

47. Environmental Action Coalition, 625 Broadway, New York, NY 10012; (212) 677-1601 or The Student Conservation Association, Inc. P.O. Box 550, Charlestown, NH 03603; (603) 543-1700.

48. The Wilderness Society, 1400 Eye Street NW, Washington DC 20005; (202) 842-3200.

49. Earth Force, 1501 Wilson Blvd, 12th Floor, Arlington VA 22209; (703) 243-7400 or CAPE, The Childrens Alliance For Protection of the Environment, P.O. Box 307, Austin, TX 78767; (512) 219-7838.

50. The Children's Earth Fund, 40 West 20th Street, 11th Floor, New York City, NY 10011; (212) 727-4505.

Suggestions For Further Reading

1. *Design For A Livable Planet: How You Can Help Clean Up The Environment* by Jon Naar, Harper and Row, New York, 1992.

2. *Earth in The Balance, Ecology and The Human Spirit,* by Al Gore, Houghton Mifflin Company, New York, 1992.

3. *The Earth Speaks* by Steve Van Matre and Bill Weiler, The Institute for Earth Education, Box 288, Warrenville, IL, 60555, 1991.

4. *Green Lifestyle Handbook,* Edited by Jeremy Rifkin, Henry Holt and Company, New York, 1990.

5. *Save Our Planet:* 750 Everyday Ways You Can Help Clean Up The Earth, by Diane MacEachern, Dell Publishing, New York, 1990.

The Teacher's Guide To The Earth Pledge

On the 20th anniversary of Earth Day in 1990, The Earth Pledge was introduced at Stanford University. Since then something curious and wonderful has happened with The Earth Pledge. It has been embraced by people of all backgrounds. They have each brought their own unique creativity and energy to the message. Here are just a few of ways that people are using the Earth Pledge in School Programs. We encourage you to either use these ideas or create your own.

✧ Start the school day with a voluntary recitation of the Earth Pledge. One school writes," Then we set aside time for the children to feel heard, to share ideas and to share concerns." Some schools then discuss Earth Stewardship. An Earth Steward is someone who: 1) takes the Earth Pledge 2) resolves to incorporate it into their daily behavior patterns 3) tells four friends about it and 4) let's us know (See form on page 160).

- ✧ Develop a partnering relationship with a school in another state or country and share information about your environmental problems and solutions.
- ✧ Have students write your local newspaper, radio station, television station or a friend from another state or country telling them about the Earth Pledge Campaign.
- ✧ Sing the Earth Pledge Song. It has been put to music both by professional musicians and by students. Several schools now sing The Earth Pledge Song in their classrooms each morning.
- ✧ Create an Earth Pledge Wall. Have students draw pictures on ceramic tiles illustrating the Earth Pledge. Glaze the tiles and mount them on a wall in a prominent place at the school. The Earth Pledge wall then creates a permanent monument not only to the words of the Earth Pledge, but also to the artwork of the students.
- ✧ Have students draw pictures illustrating the Earth Pledge and include them in an Earth Pledge art exhibit on classroom walls.

- ✧ Write essays or draw pictures (See Essay Contest form on page 146) inspired by the words of the Earth Pledge. We invite you to send student essays and pictures to us at: Earth Pledge Program: Cronkhite Beach, Building 1055, Sausalito, CA 94965. We will select the best ones and include them in *The Children's Earth Pledge Book*.

- ✧ Organize community events (concerts, lectures, fairs) around the themes of The Earth Pledge. Use each line of the pledge to spur discussion or creative expression about issues of the environment, human rights, peace and social justice.

- ✧ Invite students to go out into their neighborhoods to sell Earth Pledge Posters, Books, Postcards, and T-shirts as a way of raising money for your environmental education programs. (See Earth Pledge Product Order Form on Page 159.)

- ✧ Have your students write a letter to their favorite celebrity or their elected representatives asking them to take the Earth Pledge and declare themselves Earth Stewards.

The Earth Pledge Essay and Illustration Contest
Include entries on separate sheets and return to:
One World Inc. Cronkhite Beach, Building 1055, Sausalito, CA 94965

(Winners will receive posters and t-shirts. Their essays and
illustrations will be included in the *Children's Earth Pledge Book.)*

I. Essay Questions:
A. What is the web of life?
B. What is dignity?
C. What is the global family?
D. What does it mean to protect the earth?
E. What do the words of The Earth Pledge make you want to do?
F. What is the best way to insure lasting peace in the world?
G. What can we do to guarantee freedom and justice for all?

II. Illustrations. Please draw a picture of:
A. One planet, one people, one world in harmony.
B. People living in peace, with freedom and justice for all.
C. The web of life.

THE URGENCY OF THE EARTH PLEDGE

There is an urgency in our environmental predicament that begs for a listening. There is an instrument that is already creating that listening. There is an instrument that holds the possibility of trim tabbing the great ship of state headed full steam in a direction that must change quickly if it is to avoid disaster. The great ship can only be turned quickly by a tiny trimtab on the rudder. This tiny instrument is a powerful speaking that can change the direction of the great ship of state. This instrument, this speaking, this trimtab is the Earth Pledge.

The Earth Pledge is more than just another pledge. The Earth Pledge is an instrument of transformation. Change takes time. Transformation takes place in an instant. Change creates upset. Transformation creates hope, vision and inspiration. We cannot have true change without transformation. When you speak the Earth Pledge, you are invited to speak it with conviction from your heart.

This powerful speaking of a heartfelt commitment is the instrument we need to shift direction quickly and effect long term change. Committed speaking creates the future. The Earth Pledge's simple promise creates the listening that's begged for. It calls forth our highest selves and gives us immediate inspiration. The Earth Pledge is the instrument that provides what the urgency calls for - an opportunity to speak into existence, to create with our promise, the transformation we need to effect change, to create a future with promise. The Earth Pledge has the power to unite us, align us and empower all of us to make a difference in the future of our planet. Join us in speaking from your hearts the most important message of our time, the Earth Pledge.

• **Susan Cadogan, President, Earth Angel Inc.**

The Earth Pledge Song* (Acoustic Version) by Mark Stanley

*For children's version by F. Richmond (arranged for piano) contact: Hal Leonard
Corporation, 7777 W. Bluemound Rd., P.O. Box 13819, Milwaukee, WI 53213

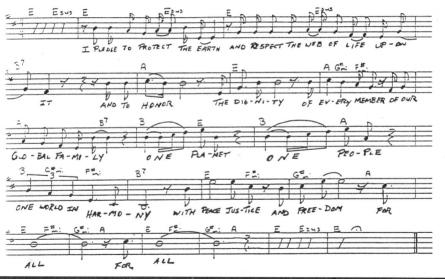

I PLEDGE TO PROTECT THE EARTH AND RESPECT THE WEB OF LIFE UP-ON

IT AND TO HONOR THE DIG-NI-TY OF EV-ERY MEMBER OF OUR

GLO-BAL FA-MI-LY ONE PLA-NET ONE PEO-PLE

ONE WORLD IN HAR-MO-NY WITH PEACE JUS-TICE AND FREE-DOM FOR

ALL FOR ALL

Ethical Principles
To Guide The Use of the Earth Pledge.

Can These Points Become Principles of Interaction Within The Classroom and Broader Community?

1. **Cooperation, Coordination, Collaboration** - By developing cooperative and collaborative relationships we seek to avoid duplication of effort, competition, waste and inefficiency. We also seek to move beyond adversarial relationships to establish trust and understanding as the primary motivators in personal interaction. The Earth Pledge is non-sectarian, non-denominational, and non-political.

2. **Inclusiveness** - Fundamental to the mission of the Earth Pledge is creating a sense of inclusiveness. People of all cultural backgrounds, political persuasions, age groups, races,

genders, nationalities, professions and religions are welcome within the community of those affiliated with The Earth Pledge.

3. **Healing and Growth** - Within the community of those associated with the Earth Pledge there is the potential for healing and ethical growth. One of the central purposes of the Earth Pledge is to provide a non-judgmental home for those who wish to find a safe space to grow and regain their centeredness.

4. **Positive Mindedness** - The Earth Pledge is about affirming our faith in the future with a sense of positive mindedness. Though our collective speaking of the Earth Pledge we seek to consciously create a future with hope and promise.

5. **Consensus Building and Commonalities** - The Earth Pledge is designed to help build broad consensus behind one simple, but powerful statement of principles. Our goal is to focus attention on our common interests and our common goals.

6. **Outreach To All Sectors of Society** - The Earth Pledge seeks to actively engage nonprofit groups in the fields of environment, peace, human rights and social justice. The Earth Pledge also seeks to reach out to all other sectors of our society including government, the religious community, secondary schools, universities, the business community, the media, the medical field, the scientific community, the entertainment community and the legal community.

7. **Enjoyment** - The Earth Pledge, and all of us associated with it, will be most effective when there is a sincere feeling of enjoyment in our efforts. To build this sense of enjoyment, humor, music, celebration, dance, sport, and spontaneity are all encouraged as part of locally inspired programs for the Earth Pledge Campaign.

8. **Multiculturalism** - Customs, rituals and ceremonies of all cultural, racial, ethnic traditions are welcome within the context of the Earth Pledge Campaign.

9. **Professionalism and Discipline**   - In spreading the message of the Earth Pledge, people should be disciplined and professional in their approach. Earth Stewards need not be stiff or rigid. In fact, spontaneity and originality are encouraged. But basic considerations of courtesy, care, attention to detail and civility should be honored at all times.

10. **Imagination and Creativity** - We seek to call into play our active imaginations and originality in creating programs involving the Earth Pledge. Whatever comes from the honest and sincere expression of self is worthy and true.

11. **Partnering** - In creating programs for spreading the message of the Earth Pledge, emphasis will be placed on ideas that build upon the respective strengths of partners and participants. It is hoped that by working in such partnerships the whole will be significantly greater than the sum of the parts.

12. **Technological Relevance** - In spreading the message of the Earth Pledge every effort should be made to design expressive forms that utilize the most recent and relevant technologies available. Computer networking, video production, CD ROM, and other new technologies should be effectively and creatively utilized.

13. **Emphasis on Children and Schools** - Children are the future, and have the most real stake in the issue of sustainability. Children bring special qualities to the world including an ability to see the world in holistic terms. In organizing programs for the Earth Pledge, emphasis will be placed upon activities that may have a special appeal to children.

14. **Celebrity Involvement** - We live in a celebrity culture. Celebrities are a source of inspiration to many people. Thus, efforts will be made to actively involve celebrities in the promotion of the Earth Pledge.

15. **Appreciating The Power of Our Words and The Importance of Keeping Promises** - The Earth Pledge is about consciously creating a future with promise through our speaking. Our committed speaking creates hope. When we learn to keep our promises we learn the meaning of integrity.

16. **Long Term Thinking** - The objectives of the Earth Pledge Campaign are long term. The ultimate vision of the Earth Pledge is to create a widespread speaking of the pledge in as many forums and situations as possible so that it becomes integral to our way of thinking and behaving. It may take many years to realize this objective. Thus, this is a long term commitment.

17. **Fiscal Responsibility** - All viable activities and programs must be built upon a stable financial foundation. Therefore, careful forethought and attention must be paid to responsible financial stewardship. This stewardship may include sale of Earth Pledge products by schools and community groups as a fundraiser for their own programs of environmental awareness.

18. **Openness, Trust and Honesty** - People associated with the Earth Pledge should feel free to openly and respectfully express their personal aspirations, ideas, beliefs, misgivings, concerns, fears and doubts. The sooner such feelings are expressed the better.

19. **Community Involvement** - Although the Earth Pledge is global in context, the basis of involvement must come from the community. Whenever possible student councils, local town councils, school boards, school committees, and state and local governments should be involved.

20. **Spiritual Consciousness** - Diverse forms of spiritualism are welcome within the community of those associated with the Earth Pledge. Native American rituals and other religious ceremonies are encouraged as a way of reaffirming our connectedness with the earth.

21. Active Involvement - The public speaking of the Earth Pledge is an important activity in and of itself. It is through our committed speaking that we create our future. But our pledges are made all the more powerful when linked with specific hands-on activities of Earth Stewardship. A natural outgrowth of the Earth Pledge Campaign is a tie in with finite projects of Earth CPR (Conservation, Protection and Restoration). Therefore, included with fifty of the pledges in this book there is a contact number with a corresponding address listed on page 136 of this book. Please get actively involved by contacting the appropriate organization to learn what you can do.

Earth Score: Your Personal Environmental Audit & Guide

"If you're trying to live a little greener, and you're not sure how you're doing, you need EarthScore." **National Wildlife Federation**

"EarthScore allows you to chart your planetary impact and offers sensible tips on boosting your rating." **Sierra Magazine**

"Preserving our planet for humans and other species depends on the decisions we all make as consumers every day, every hour. I know of no better way each of us can evaluate our decisions than the method devised by Don Lotter in EarthScore." **Harold Gilliam, San Francisco Chronicle**

8 1/2 x 11 book, includes a 11 x 17 chart that will help you track your "Impacts" and "Actions" on the earth plus resources and tips that will help you become more earth gentle. Send check or money order for $4.50 to Morning Sun Press, PO Box 413, Lafayette, CA 94549

Fundraising? Sell Earth Pledge Products!

Would you like to raise funds for your school, church, or community group and help spread the message of The Earth Pledge?

If so, please return this order form.

(To receive wholesale prices, minimum order: $25)

Item	Retail	Your Cost	You Keep	# Items X Your Cost = Total Cost
Books	$6.95	$3.45	$3.50	___/_____
Posters	$8.00	$4.00	$4.00	___/_____
Postcards	$.50	$ .25	$ .25	___/_____
Bumpersticker	$1.00	$ .50	$ .50	___/_____
T-shirts	$14.00	$7.00	$7.00	s_m_l_xl__/_____
		Grand	Total	_____

To receive an "Earth Pledge Starter Kit"* return this form along with $25.
* Includes 20 postcards, 1 book, 1 poster, 1 bumpersticker & 1 T-shirt

Please make check payable to "The Earth Pledge" and send to:

The Earth Pledge Campaign, Cronkhite Beach, Building 1055, Sausalito, CA 94965

Or use Visa or Mastercard #_____ _____ _____ _____Expiration __ __

I have taken the Earth Pledge and declared myself an Earth Steward.

Name:_____ Age:_____

School:_____ Grade:_____

Home Address: _____

City, State, Zip:_____ Phone:(   )_____

My special Earth Pledge is: _____

_____

_____

_____

_____

_____

_____

_____

By returning this form, I give my permission to use my special Earth Pledge as part of a book
Clip out this page and send it to us at the **Earth Pledge Program**:
Cronkhite Beach, Building 1055, Sausalito, CA 94965.